AF694096

TimeTracks

A JOURNEY THROUGH THE NATURAL WORLD AT THE

NEW MEXICO MUSEUM OF NATURAL HISTORY & SCIENCE

COVER: The *Coelophysis* dinosaur was a swift and nimble predator that roamed New Mexico about 225 million years ago. *Coelophysis* is the official state fossil of New Mexico, and the Museum's life-sized bronze by New Mexico artist Phillip Bircheff graces the Atrium. Photo by John Arnold, NMMNHS.

Copyright © 2001 by New Mexico Museum of Natural History and Science.

All rights reserved, including the right to reproduce this work in any form whatsoever without permission in writing from the publisher, except for brief passages in connection with a review.

For information, write:

The Donning Company Publishers, 184 Business Park Drive, Suite 206, Virginia Beach, VA 23462
Steve Mull, General Manager
Ed Williams, Project Director
Debbie Williams, Project Research Coordinator
Dawn V. Kofroth, Assistant General Manager
Sally Clarke Davis, Editor
Scott Rule, Graphic Designer
Scott Rule, Senior Marketing Coordinator
Patricia Peterson, Marketing Assistant

Library of Congress Cataloging-in-Publication Data

TimeTracks: a journey through the natural world at the New Mexico Museum of Natural History and Science.
p. cm.
ISBN 1-57864-135-7 (pbk. : alk. paper)
1. Natural history–New Mexico.
2. Historical geology–New Mexico. I. New Mexico Museum of Natural History and Science.
QH105.N6 T56 2001
508.789–dc21

2001028083

Printed in the United States of America

1–57864–135–7

Above: Apollo 16 photograph of Earth centered on New Mexico during trans-lunar injection. NASA.

"Nothing is permanent except change."
—Heraclitus

Deep within the Earth, molten rock churns and seethes, occasionally erupting on the surface. Continents are carried about on wandering plates of the Earth's surface that collide or are ripped apart. The forces of wind and water sculpt the land. Within the oceans and on the land, life is everywhere, constantly changing, with new forms arising as others disappear. The thin, fragile veneer of life, protected by the envelope of atmosphere, is only part of the living, evolving planet Earth.

At the New Mexico Museum of Natural History and Science, visitors discover the fascinating story of Earth's natural history in the permanent exhibits theme, *TimeTracks*. From Earth's very beginnings to New Mexico today, visitors experience the interwoven fabric of geological and biological processes and patterns.

As you read this book, you are about to embark on a journey through time. For this journey, you will need to think in terms of not hundreds or thousands of years, but of millions and billions of years. How can we comprehend so grand a scale? One might consider the insignificant period of human civilization when compared to the age of the Earth: if all of Earth's time were represented on a 24-hour clock, the entire history of civilization would take up only the last few seconds.

In your journey, you will race through the first 22 hours of this 24-hour clock of Earth's history, time ticking in billions, then hundreds of millions of years. The ticking will gradually slow to millions, hundreds of thousands, then thousands of years as you return to the present. Of course, the clock never stops: the processes of change, catastrophic or subtle, continue. The tracks of time, guiding you on your journey, are sometimes distinct, sometimes nearly imperceptible. Fresh time tracks are always being made, there for the curious to follow.

Your journey begins about 4.6 billion years ago, with the formation of the solar system. Bon voyage!

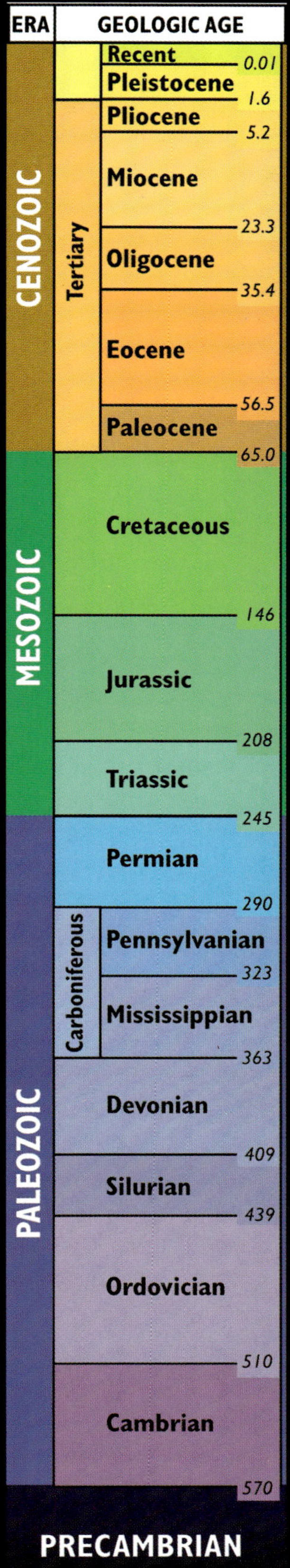

Geologic time scale: The time before the present is shown in millions of years. Image by Matt Celeskey, NMMNHS.

As our solar nebula became denser, it flattened out into a disk like a spinning pizza crust. A central spherical concentration within this disk grew rapidly, and its temperature rose so high that thermonuclear reactions were started, and a star, our Sun, was formed. In the surrounding, rotating disk of gas and dust, the dust particles massed together into larger and larger bodies that orbited the Sun, eventually forming the planets. Concentrations of chemical elements, and the minerals formed by these elements, varied depending upon the distance of the planet from the sun.

Asteroids, dust-sized to kilometer-sized chunks of rock, are believed to be remnants of the formation of our solar system. Asteroids orbit between Mars and Jupiter, and apparently have not formed a planet because of the interfering gravitational pull of the large planet Jupiter. Some asteroids fall from their orbit, hitting Earth as meteorites. Tons of dust-sized debris rain down on Earth each day. The oldest meteorites, which scientists think were formed at the same time as Earth and the other planets, date the formation of our solar system at about 4.6 billion years ago.

Comets, other remnants of the formation of our solar system, orbit about the solar system's cold outer reaches, as far as six trillion miles away. Made of ice and dust, comets are often called dirty snowballs. As they approach the Sun, the ice turns to gas, and vapor and dust stream outward to form the comet's tail. Some comets may take 100,000 years to complete their orbit about the Sun, while others take only three years. People on Earth can see Halley's comet every 76 years.

Bombarded by meteorites and comets, early Earth probably looked much like the moon. Heat from these impacts and from the radioactive decay of unstable elements melted some of the accumulated rock, and the heavy material sank, while the lighter material rose. The heavy material (mostly iron) formed Earth's core; the lighter material (mostly oxides and silicates) formed the seeds of our present-day continents. The rocks of Earth's interior were in constant motion as the hotter material rose, shedding their heat to the surface and sinking as they cooled. Volcanoes erupting on the surface released gasses that became the early atmosphere and condensed to form the oceans. Meteorites and comets also delivered much of Earth's water during the early bombardment of the planet.

Great masses of land moved across Earth's active surface. Rocks were brought to the surface along mid-ocean rifts and crushed back deep into the Earth along the edges of colliding surface plates. This constant resurfacing of the planet has erased much of the early record of Earth's history.

For millions of years, the dynamic, restless Earth was devoid of life. Energy from early volcanoes, lightning, and intense ultraviolet sunlight turned the oceans and atmosphere into a chemical laboratory, brewing new chemical compounds from nitrogen, methane, ammonia, carbon dioxide, hydrogen, and water. After millions of years, this brew became an organic soup of alcohols, fats, carbohydrates, amino acids, and larger molecules similar to proteins.

The appearance of one type of molecule in the oceans changed the random and sporadic production of these organic molecules. This molecule, a nucleic acid, could reproduce itself, carrying the code for replication in its structure. This code also assembled building blocks of amino acids into proteins, which became protective membranes and accessory structures around the nucleic acid. The nucleic acid (deoxyribonucleic acid, or DNA) thus acted as the control center for a self-replicating, self-sufficient "package"—the first cell.

Unicellular Life: Variations on a Theme

An aquarium in *Origins*: Before 400 million years ago, all life on Earth lived in the seas. Photo by David Mandel, NMMNHS.

The earliest living things were simple, one-celled packages much like modern bacteria. They grew and survived by consuming organic materials in the oceans. Although very simple, bacteria are still Earth's most abundant organisms, inhabiting every conceivable nook and cranny on Earth. They grow in deep-ocean volcanic vents at temperatures of 400°C and in Arctic ice well below 0°C. Bacteria are critical links in every food chain on Earth. Without bacteria, all other life on Earth would quickly disappear.

About 3.5 billion years ago a type of bacterium that did not depend on the organic soup for food emerged. It could turn carbon dioxide gas into food sugars using energy from sunlight, a process called *photosynthesis*.

Early photosynthetic organisms have been preserved in fossils called *stromatolites*. *Stromatolites* (meaning "stone blankets") are mounds built of sediment trapped by dense layers of blue-green bacteria. New layers of bacteria grew on the sediment layers, forming dome-like mats. Fossil *stromatolites* 3.5 billion years old provide the earliest record of cellular life.

Although living stromatolites exist today only in a few isolated tidal zones of tropical oceans, they blanketed Earth for more than two billion years. *Stromatolites* produced great amounts of oxygen as a by-product of their *photosynthesis*, and in so doing completely changed the composition of the atmosphere. This excess oxygen combined with iron in the oceans to precipitate great beds of iron oxides. It also formed an ozone layer in the upper atmosphere, shielding Earth's surface from much of the harmful ultraviolet sunlight. The abundant oxygen poisoned many organisms, but some could use this oxygen to power their cells. Most modern life descended from these first oxygen users.

Another, more complex kind of cell arose about 1.5 billion years ago. In these new cells, different compartments performed specific tasks. The nucleus contained the genetic code (DNA), which controlled the growth and production of the cell. Two kinds of compartments are descendants of bacteria that invaded or were engulfed by the cell, but became interdependent parts of it: mitochondria, powerhouses of animal and plant cells; and chloroplasts, which build sugar from sunlight, carbon dioxide, and water in plants. These more complex one-celled creatures quickly spread and diversified to occupy freshwater and marine habitats.

Numerous meteorites of all types and sizes have been found on Earth. The easiest ones to find are the metallic meteorites because they look very different from Earth rocks. A metallic meteorite estimated to have been 90 feet in diameter, weighing 20,000 tons, and traveling with a velocity of 15 miles a second, formed Meteor Crater in Arizona.

Multicellular Life: A Society of Cells

In the struggle for life in the oceans, it was a cell-eat-cell world. Those cells with protective coverings or defense mechanisms fared better. Size alone provided a great advantage: it was difficult for a smaller cell to engulf a larger one. Cells clinging together in a colony were afforded even greater protection.

In a colony, different cells could specialize for different tasks, such as defense or reproduction, making a more efficient organism. "Societies" of interdependent cells that reproduced as units became the different species of multicellular life. From these early multicellular organisms came the myriad forms that spread through the oceans of the world: shelled creatures, kelp, sea urchins, corals, crabs, and the first vertebrates. Among these were the future colonists of dry land.

Fossil Crinoids. Although they look like plants, these late-Paleozoic invertebrates were animals that attached themselves to the ocean bottom by a stem. Most modern crinoids do not have stems. The limestone on top of Sandia Crest overlooking Albuquerque once was a Paleozoic ocean; it has fossil crinoids. Photo by David Mandel, NMMNHS.

Success on Land

Four hundred million years ago, during the Paleozoic ("early life") Era, a few aquatic plants established a tentative foothold on some swampy shoreline, and the first terrestrial animals followed soon after. Within 100 million years, terrestrial plants and animals had conquered the problems of drying out, obtaining nutrients, and locomotion on land, and they thrived in moist lowland regions of Earth, although still requiring water for reproduction. Plants first broke fully away from water. Gymnosperms (conifers and

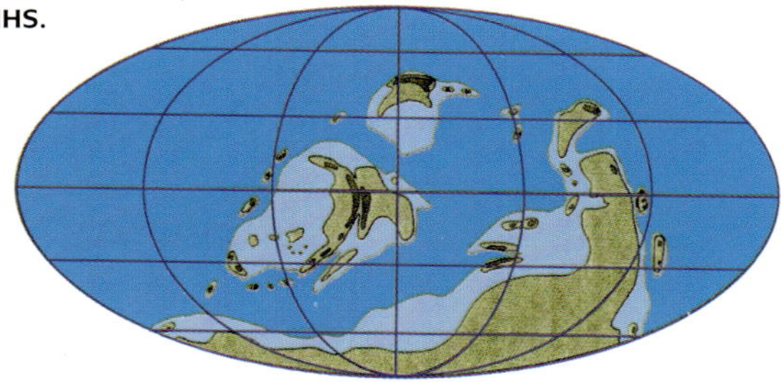

Distribution of land (light brown), mountains (dark brown), shallow seas (light blue), and ocean basins (dark blue) 420 million years ago.

Muscle tissue can be seen in this finely preserved amphibian fossil from Pennsylvanian-age (300 million years ago) deposits in the Manzanita Mountains east of Albuquerque. Photo by Spencer Lucas, NMMNHS.

Early Permian red-beds in the Robledo Mountains of southern New Mexico preserve vertebrate and invertebrate trackways that allow a detailed reconstruction of life 280 million years ago. Photo by Spencer Lucas, NMMNHS.

Fossil conifer, preserved in the Big Hatchet Mountains of southern New Mexico. Gymnosperms were among the first plants to live on land. Photo by Spencer Lucas, NMMNHS.

relatives) and seed ferns produced air-born pollen to fertilize distant eggs, while the developing embryo was released from the parent in a protected package—the seed. Shortly thereafter, reptiles developed an egg that could develop out of the water—the amniote egg. In this egg, the embryo develops in its own tiny pool of fluid, protected from the outside world by enveloping layers of membrane that also provide food, waste disposal, and respiration.

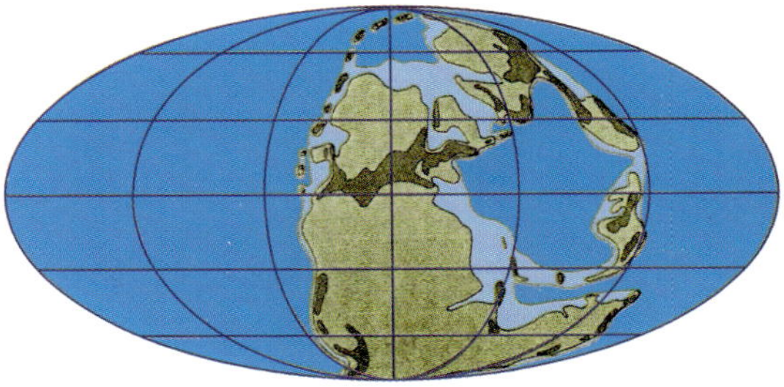

Distribution of land (light brown), mountains (dark brown), shallow seas (light blue), and ocean basins (dark blue) 250 million years ago.

By 250 million years ago, most of the landmasses had converged into a supercontinent, Pangaea. Massive glaciers locked much of the oceanic water in ice, exposing land along continental margins. The seas that had covered most of New Mexico withdrew, and the extensive shallow-water reefs that had grown for 20 million years declined. The marine life that lived in these Paleozoic seas can be seen today as fossils in the limestone at the top of the Sandia Mountains, overlooking Albuquerque to the east. Perhaps because of these global climatic changes, life suffered a catastrophic blow at the end of the Paleozoic Era, about 245 million years ago: nearly 90 percent of life forms died away. This marked a major change in life on our planet, and opened the way for the dominance of the reptiles and dinosaurs.

Red sandstone of the Permian Period (290 to 245 million years ago) is visible in the lower canyon walls near Jemez Springs, New Mexico. The sandstone has been covered by tan-colored Bandelier tuff that erupted from the Valles Caldera about one million years ago. Photo by L. S. Crumpler, NMMNHS.

Dawn of the Dinosaurs

The Age of Reptiles began 245 million years ago in a world dominated by a single supercontinent—Pangaea. Plants and animals thrived on land, no longer restricted to the oceans in which life had evolved for more than three billion years.

New Mexico and southwestern North America were part of an immense flood plain of subtropical forests, through which rivers flowed on their way to the western seas. Mosses, ferns, horsetails, and tree ferns grew in wet lowlands, while primitive conifers grew to ninety feet tall in the drier

Many fossil skeletons of the early dinosaur *Coelophysis* (New Mexico's state fossil) were found at Ghost Ranch, New Mexico. *Coelophysis* is shown here in its early Triassic environment (the Museum's *Dawn of the Dinosaurs* mural is by Margaret Colbert). Photo by David Mandel, NMMNHS.

FUN FACT

Look closely at the Margaret Colbert mural in the "Dawn of Dinosaurs" exhibit: Only one true dinosaur—*Coelophysis*, the New Mexico state fossil—is represented as the small striped biped. One small mammal also is in the mural: this time period was the beginning of both dinosaurs and mammals, but mammals remained small in size until after the extinction of the dinosaurs.

uplands. A rich diversity of animals lived in the warm, moist forests and numerous rivers and streams.

Dinosaurs and mammals, both destined to become dominant animals on land, first appeared in the Triassic Period, more than 200 million years ago. Small reptiles and mammals fed on primitive insects (dragonflies and beetles). The early dinosaur *Coelophysis* (New Mexico's state fossil) was a nimble predator that lived in herds or packs. Clams and freshwater horseshoe crabs lived in the shallow waters, along with the last of the large amphibians, the metoposaurs. Freshwater sharks and lungfish swam in the streams and marshes, pursued by the giant, crocodile-like phytosaurs.

The rocks that preserve the record of the Triassic environment form colorful landscapes in today's Southwest: the Painted Desert and Petrified Forest of Arizona, and the red mesas and canyon lands of northern New Mexico.

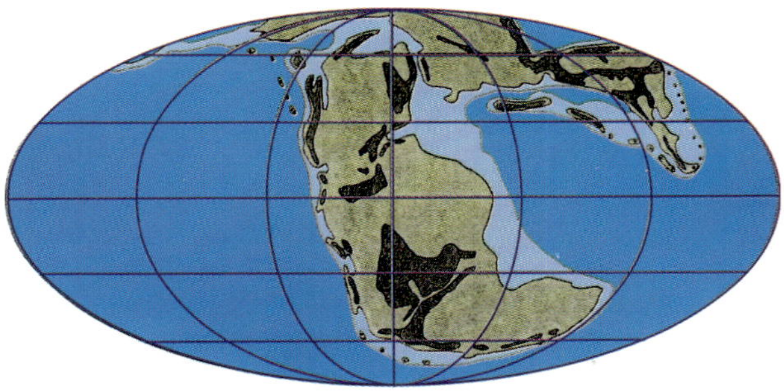

Distribution of land (light brown), mountains (dark brown), shallow seas (light blue), and ocean basins (dark blue) 210 million years ago.

New Mexico artist Phillip Bircheff based his bronze reconstruction of *Coelophysis* (in the Museum's Atrium) on fossilized specimens found near Ghost Ranch, in northern New Mexico. Photo by Spencer Lucas, NMMNHS.

Triassic (lower) and Jurassic (upper) strata, such as these near Ghost Ranch, provide colorful and dramatic vistas of northern New Mexico. Photo by Spencer Lucas, NMMNHS.

Age of Giants

The Age of Giants began about 200 million years ago when North America was still part of the supercontinent Pangaea. As Pangaea broke apart during the next 60 million years, North America drifted northward away from the equator, dramatically changing New Mexico's environment.

In New Mexico and the Southwest, layers of sedimentary rocks have left a vivid record of Jurassic time. The massive sandstones of the Entrada Formation were formed from a great sea of sand that covered southwestern North American 175 million years ago and was very similar to today's Sahara Desert of northern Africa. Extensive dunes became buff-colored cliffs of sandstone. No record of the plants or animals that lived in or near this great sea of sand exists today. Elsewhere, dinosaurs and giant marine reptiles dominated the lands and the seas.

Later, a shallow sea extended from southern Canada in a narrow band that almost cut North America in two. An arm of this sea became an isolated saline lake (similar to today's Great Salt Lake in Utah). As the lake dried, limestone and gypsum were deposited. These deposits make up the

To create the mural in the *Age of Giants*, Canadian artist Ely Kish sculpted accurate models of dinosaur skeletons in their eventual pose and added clay muscles and skin. She then traced projected images of the models on the wall before painting the mural. Photo by David Mandel, NMMNHS.

Extensive dunes of the Entrada Formation are preserved in colorful cliffs of northern New Mexico. Cross bedding of the original dunes often is visible, as in this outcrop near Clayton, New Mexico, and on a touchable specimen in *Age of Giants*. Photo by Spencer Lucas, NMMNHS.

Todilto Formation, which is quarried for gypsum in New Mexico today.

After the shallow sea had dried, rivers and forests covered the land. Warm climate and abundant food fostered the development of dinosaurs, the largest animals ever to live on land. In New Mexico, *Allosaurus*, *Stegosaurus*, and *Camarasaurus* lived amidst the rivers and forests, now preserved in mudstones and sandstones of the Morrison Formation. The Morrison Formation extends all the way to Montana, and frequently is the site of fossil dinosaur discoveries.

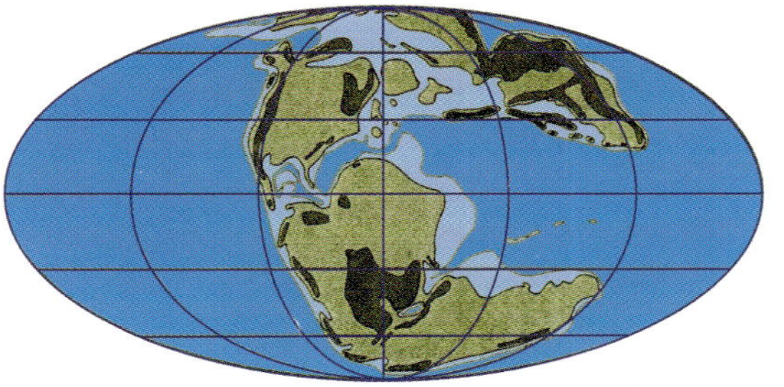

Distribution of land (light brown), mountains (dark brown), shallow seas (light blue), and ocean basins (dark blue) 160 million years ago.

Two very different flying animals lived during the Age of Giants: pterosaurs, or "winged lizards," and the first birds. Distantly related, these two flying forms coexisted until about 65 million years ago, when the pterosaurs disappeared. Modern birds, descendants of the primitive *Archaeopteryx*, have diversified into a vast array of sizes, shapes, and habits.

The Zuni sandstone, deposited during the Jurassic Period, catches the late afternoon sun near Grants, New Mexico. This sandstone, which is hundreds of feet thick, originated as wind-blown sand in the Sahara-like sand sea that covered most of New Mexico about 170 million years ago. Photo by J. C. Aubele.

FUN FACT

Fossilized dinosaur bone is extremely heavy; most mounted skeletons in museums are lightweight casts of the bones. The casts are painted black in this exhibit because most of the original bones came from the Cleveland-Lloyd quarry in Utah: bones from this quarry are dark in color due to trace elements in the rock from which they are excavated.

Casts of *Allosaurus* and *Stegosaurus* from Utah, stand before a late Jurassic scene in New Mexico, 160 million years ago (mural in *Age of Giants* by Ely Kish). Photo by David Mandel, NMMNHS.

The earliest pterosaurs (flying reptiles), like these rhamphorhycoids soaring above the *Age of Giants*, had long tails and uncrested heads. Later pterosaurs had shorter tails and crested heads. Photo by David Mandel, NMMNHS.

FUN FACT

Canadian artist Ely Kish painted the mural in the Age of Giants (shown on pages 14 and 15). To create the dinosaurs in the mural, Kish first sculpted accurate models of their skeletons in the eventual pose and added clay muscles and skin. She then traced projected images of the models on the wall before painting the mural (in place) from scaffolding.

New Mexico's Seacoast

During the Cretaceous Period (144–65 million years ago), the world was a warmer and wetter place than today. Vast, shallow continental seas covered at least one-third of today's land area. These extensive oceans absorbed the sun's heat, spreading it uniformly around the world. Polar ice caps did not exist, and dinosaurs lived north of the Arctic Circle.

About 100 million years ago, a seaway that extended from the Gulf of Mexico through Canada to the Arctic Ocean covered New Mexico. Ammonoids (coiled, shelled animals up to a meter across that were the ancestors of today's nautilus) swam in these shallow seas, along with sharks and marine reptiles (giant

plesiosaurs and mosasaurs). As the seaway waxed and waned, its shoreline shifted across northern and central New Mexico in a series of northwest-to-southeast bands. The rocks of the San Juan Basin in northwestern New Mexico record these fluctuations in alternating layers of marine shale deposits, and swamp and river sediments. At the end of this Period, crustal plates slammed into the western margin of North America, compressing the continent and thrusting up the modern Rocky Mountains as far south as the Sangre de Cristo Mountains of New Mexico. This uplift expelled the seas from the continent near the end of the Cretaceous, and changed the climate and landscape of western North America.

Much of the Cretaceous shoreline swamp and forest vegetation has been transformed into the rich coal, oil, and gas deposits of northern New Mexico. Well-preserved leaves and petrified logs in these deposits tell us a great deal about the ancient shoreline environments. Along the shorelines grew lush forests of plants, much like modern species.

The Bisti badlands of the San Juan Basin, in northwest New Mexico, preserve a rich record of the plants and animals that lived in swamp forests along a shallow inland sea 100 million years ago. Photo by Spencer Lucas, NMMNHS.

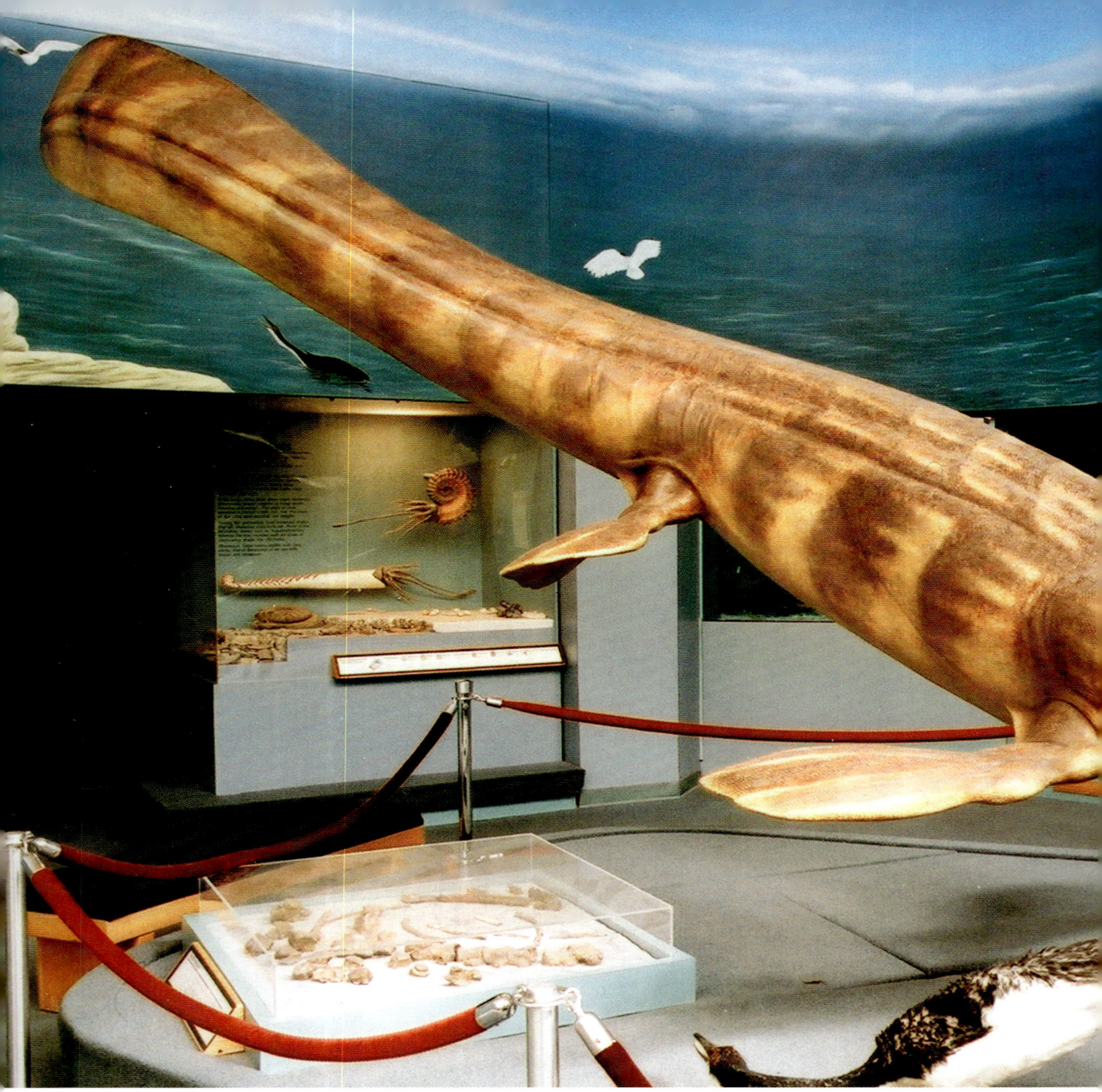

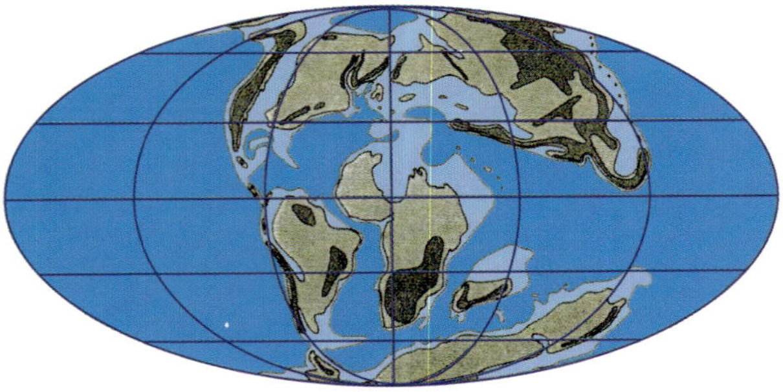

Distribution of land (light brown), mountains (dark brown), shallow seas (light blue), and ocean basins (dark blue) 100 million years ago.

These floodplains were very similar to the Mississippi River delta and coastal areas along the Gulf of Mexico today. Areas of higher ground away from the coast were like the forests in the southeastern United States or Central America today. In the swampy shoreline forest lived crocodiles, garfish, and turtles, much as in today's Florida

Marine lizards like this mosasaur, reconstructed in *New Mexico's Seacoast*, were powerful and efficient predators up to thirty feet long that lived throughout Earth's seas during the Cretaceous Period. Photo by David Mandel, NMMNHS.

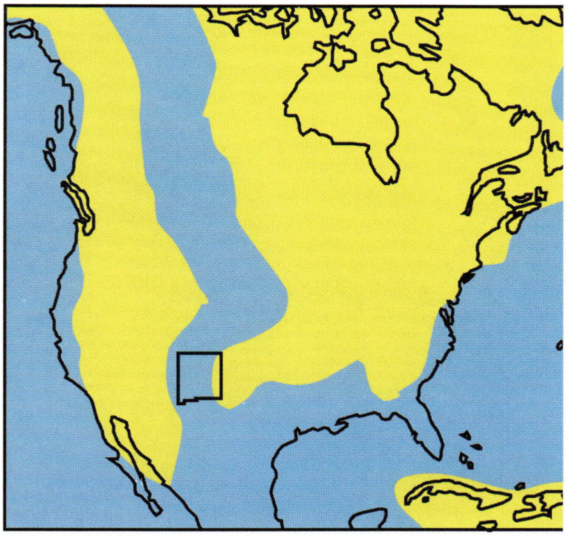

During most of the Cretaceous Period, central North America was covered by a shallow sea that stretched from the Gulf of Mexico to the Arctic Ocean. An interactive exhibit in *New Mexico's Seacoast* shows how repeated Cretaceous seaways swept across New Mexico.

FUN FACT

Tyrannosaurus rex lived during the end of the Cretaceous period, not during the Jurassic. The life-sized, bronze dinosaurs at the Museum entrance –*Albertosaurus* (the sculpture is called "Alberta") and *Pentaceratops* (affectionately known as "Spike")—also lived during the Cretaceous. *Pentaceratops* is a New Mexico native; its bones have been found only in New Mexico.

Everglades. Wandering through these forests were the giant dinosaurs, especially herbivorous duck-billed dinosaurs (hadrosaurs) and horned dinosaurs (ceratopsians), and the carnivorous *Tyrannosaurus*. As paleontologists sift deposits in the San Juan Basin, more and more traces of small but diverse mammals are being found. These were

Scientists believe that the hadrosaurs, herbivorous "duck-billed" dinosaurs like these in *New Mexico's Seacoast*, were attentive parents. Photo by David Mandel, NMMNHS.

FUN FACT

Southwestern New Mexico was at the edge of a continent during the Cretaceous, and large volcanoes similar to today's Cascade Mountains were erupting. Today, the deep cores of those Cretaceous volcanoes are mined for copper in southwestern New Mexico.

the predecessors of the modern mammals that would inherit the world from the dinosaurs.

Flowering plants (angiosperms) may have first appeared as early as 210 million years ago. For millions of years they remained a small part of a landscape dominated by ferns and non-flowering trees. Then, 70 million years ago, flowering plants greatly increased in number and variety. Today, 85 percent of land plant species are angiosperms. Rich fossil beds document this change in New Mexico.

Flowering plants are successful because they complete their reproductive cycles quickly. This allows them to survive in harsh environments where growing seasons are short. The plants also evolved a system of cooperation with animals. They produce seeds surrounded by a nutritious fruit that attracts animals. When animals eat the fruit, they disperse the seeds to new locations. Many plants have showy flowers that attract animals to feed on the sweet nectar or the pollen. In the process, the animals efficiently transfer pollen from one flower to the next. As flowering plants spread and diversified, so did the insects, birds, and small mammals that depended upon them for food.

At the close of the Cretaceous, vast shallow seas that had covered the continents began to withdraw as deepening ocean basins drained the waters from the land. The world's climate changed dramatically, as movement of Earth's surface plates carried the continents towards the poles. Heat was not transferred readily to the more polar landmasses, and zones of distinct climate developed. Greenland and Antarctica were to acquire ice caps that exist to this day. Many animals disappeared forever; dinosaurs, flying reptiles, giant marine reptiles, and the ammonoids were gone by the end of the Cretaceous.

Although major extinctions occurred during the Late Cretaceous, this time should be regarded as one of turnover, not just extinctions. Many dominant and long-lived groups did disappear: about 75 percent of all of the species on Earth became extinct. At the same time, other major groups appeared or greatly diversified (marsupial and placental mammals, flowering plants, and many new forms of insects). Many theories have been proposed to account for this floral and faunal turnover. Certainly, the drastic change in global climate was involved to some degree. Perhaps other, more catastrophic, events were also involved. Evidence of a 108-mile impact crater on the edge of the Yucatan Peninsula of Mexico shows that a giant asteroid-sized meteorite struck Earth at this time. This large impact would have thrown dust into the highest parts of the atmosphere, creating a lengthy winter that spelled doom for many life forms.

The golden sandstone and shale of the Rio Puerco Valley in New Mexico began as sand and silt deposited by a shallow Cretaceous sea. Today, the Rio Puerco has eroded much of the original rock into sand and silt and has formed mesas and cliffs. Photo by J. C. Aubele, NMMNHS.

Age of Volcanoes and Evolving Grasslands

Sixty-five million years ago, New Mexico emerged from the sea for the last time. At the time, New Mexico was warm and humid: plant fossils from the Raton Formation of northeastern New Mexico record a warm, deciduous forest with magnolia, fig, breadfruit, and cinnamon trees. Mammal fossils found in the Nacimiento, San Jose, Raton, and Galisteo formations include primitive carnivores and ancestors of hooved mammals; the rodent-like (but egg-laying) multituberculates, which survived for some 100 million years; and primitive primates.

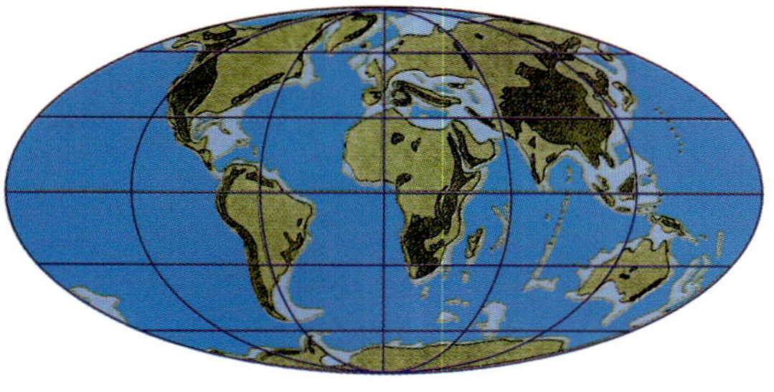

Distribution of land (light brown), mountains (dark brown), shallow seas (light blue), and ocean basins (dark blue) 15 million years ago.

Most of the continents continued to drift north as passive riders on the Earth's crustal plates, and much of the world's land mass became crowded into the northern hemisphere. Ocean currents that had once cycled from the equator to the North Pole were now blocked, and cold polar water became trapped in the north. As Antarctica drifted over the South Pole, polar ice caps began to grow at both poles. Zones of tropical, temperate, and arctic climates developed.

On the west coast of North America, chains of

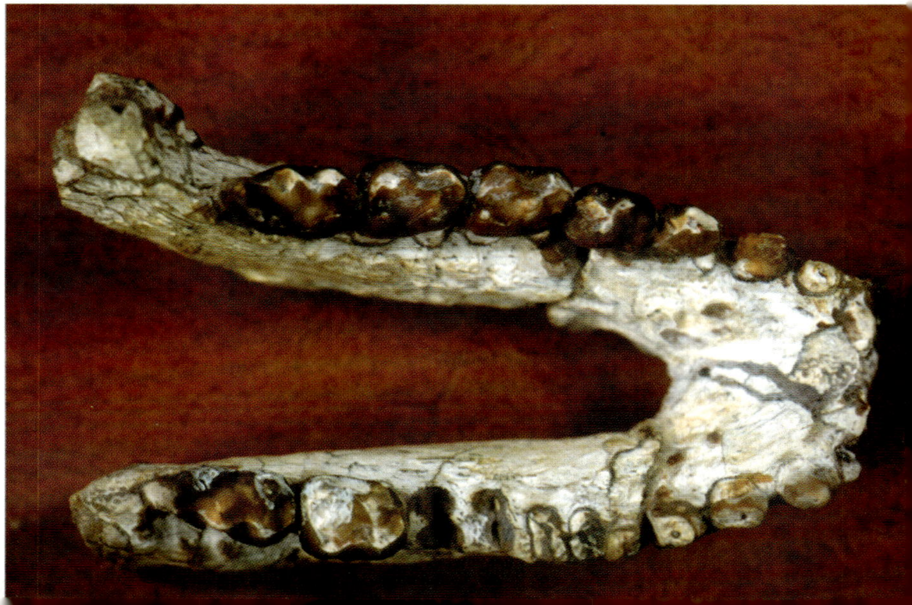

This lower jaw of an early primate, part of the Museum's extensive fossil collections, was found among the diversity of mammals preserved in 50 million-year-old rocks near Socorro, in central New Mexico. Photo by Spencer Lucas, NMMNHS.

volcanoes erupted at the plate edges and the shallow continental seas slowly drained. New mountain ranges pushed up when Earth's crust folded under the pressure of colliding plates. These mountains prevented the flow of warm, moist oceanic winds into the continental interiors. Inland areas dried out and their temperatures became more extreme. New Mexico and much of central North America began a change to cooler and drier weather. Around the world, forests of large, lush flowering plants dwindled. Grasses, which are also flowering plants, spread from marshes and streamsides to become the dominant plants of newly forming prairies.

Miocene-age badlands such as these near Española, New Mexico, preserve a rich record of the plants and animals that lived in New Mexico 15 million years ago. Fossils of this period make up an important part of the Museum's collections. Photo by Spencer Lucas, NMMNHS.

Grasses are tough, versatile survivors. Grasses can survive freezing cold, fire, drought, and lawn mowers. They grow from both aboveground and underground centers called meristems. Leaves and stems die back when

Mt. Taylor, a composite volcano in western New Mexico, last erupted about two million years ago. Its profile shows a classic cone-shaped volcano with a missing top. Photo by L. S. Crumpler, NMMNHS.

conditions are severe, when fire sweeps through, or when animals graze on them. But the leaves can sprout again from the protected underground meristem. Old, dead leaves fuel range fires that destroy woody trees and shrubs (potential

FUN FACT

Horses, camels, large cats, and several varieties of elephants all lived in New Mexico during the Tertiary Period. Their bones and tracks have been found in many areas in central and north-central New Mexico.

Hyracotherium, or the "dawn horse," was a dog-sized horse that lived in dense forests of the early Tertiary, 50 million years ago (mural in *Volcanoes and Evolving Grassland* by Jay Matternes, on loan from the Smithsonian Institution). Photo by David Mandel, NMMNHS.

competitors). Grasses rely on the wind to spread their pollen, rather than on insects. This is an efficient way to pollinate on windy plains with large, open expanses.

The expanding grasslands became proving grounds for the tolerance of harsh conditions. Those animals able to adapt to the new conditions became more numerous. Grassland specialists evolved that were able to thrive in the harsher climate and open landscape. The variety of mammals increased greatly, particularly in three groups. Large mammals that grazed on grass or browsed on trees and shrubs were well insulated from extreme temperatures. Rodents burrowed underground to escape severe conditions. Savannah primates, including our human ancestors, formed social groups for mutual protection.

At the same time that grasslands were developing, new landforms also were being formed. Between 40 and 20 million years ago, a broad arc of volcanism swept across western North America. Large volcanoes in

One of the youngest large lava flows in the continent erupted about 3,000 years ago south of Grants, New Mexico in an area known as El Malpais. The rough black lava flowed up against adjacent sandstone mesas. Photo by L.S. Crumpler, NMMNHS.

Arizona, Colorado, and New Mexico had huge eruptions that were more extensive and violent than humans have ever witnessed. Much of New Mexico's mineral wealth, including gold, silver, and copper, was redistributed and concentrated by this volcanic activity. As the large, old volcanoes in southwestern New Mexico eroded, the concentrations of economic minerals were discovered and are mined today.

FUN FACT

If you could build a box that enclosed Albuquerque, the floor of that box would have been covered by ash that erupted from Mount St. Helens in 1980. But the entire box would have been filled nearly 6,000 feet (higher than the Sandia Mountains) by ash that erupted from New Mexico's Jemez Mountains one million years ago.

About 20 million years ago, as crustal plates changed direction along the

This eroded interior of a volcano, known as Cabezon Peak, is a volcanic neck. The volcano erupted about two million years ago and was eroded by the Rio Puerco. The Rio Puerco Valley contains one of the largest concentrations of volcanic necks in the world. Photo by L. S. Crumpler, NMMNHS.

The modern river channel of the Rio Grande follows the rift valley. The Sandia Mountains form the eastern margin of the rift at Albuquerque. Photo by J. C. Aubele, NMMNHS.

west coast of North America, the entire Southwest began to stretch apart. As the land began to thin and pull apart, a linear chain of basins began to form down the center of New Mexico: the Rio Grande Rift. As the rift widened and deepened, the newborn Rio Grande began to flow through it, filling it with sediment. The margins of the rift lifted to form a range of mountains: the Sandias, Manzanos, and Los Piños. Volcanoes erupted from the cracks formed by rifting.

As the rift continued to widen and deepen, the Rio Grande continued to deposit sediments interlayered with volcanic eruptions: as much as 20,000 feet of sediment is currently beneath Albuquerque. This sediment, rich in mammal fossils, formed the Santa Fe Formation. Rabbits, beavers, shrews, and hedgehogs lived in woodlands and grasslands, along with gomphotheres (primitive elephants), rhinoceros-like *Teleoceros*, pig-like oreodonts, primitive deer, antelope, ground sloths, and many kinds of horses and camels.

New Mexico's Ice Age

During the Pleistocene Epoch (1.6 million to 10,000 years ago), at least four waves of glaciers advanced southward across North America. During glacial intervals, the climate in the Southwest was cool and moist, and isolated glaciers formed in the high mountains. Humans began to populate

This section of the *Panorama of Time* (a mural by Kent Pendleton in *New Mexico's Ice Age*) depicts the east side of the Sangre de Cristo Mountains of northern New Mexico during the last glacial stage 20,000 years ago. Photo by David Mandel, NMMNHS.

Earth, and to affect its natural history. Many large animal species evolved during the Pleistocene, only to become extinct by the end of the epoch.

Pollen from lake sediments, fossils from ancient caves, and the peculiar habits of a small rodent, the packrat, all provide valuable windows into the past, allowing us to reconstruct the Ice Age environment. Pollen shed from Ice Age plants accumulated in nearby lakes. The layers of pollen that settled on the lake bottom yield a serial history of the plants that occupied the region.

The Museum's walk-through cave exhibit, shown here, introduces you to the underground world of caves: How caves are formed, how cave formations grow, why fossils accumulate in caves, and what animals live in caves. Photo by David Mandel, NMMNHS.

Caves such as Carlsbad Caverns formed in limestone and fossil reefs of southern New Mexico. These reefs were originally deposited in Paleozoic seas. Caves are good sources of Ice Age fossils, as unwary creatures often stumble into caves or pits. Layers of bones and debris accumulate over time, recording the animals and plants from each time interval.

Packrats collect practically anything they can carry into their nests, called middens. This debris becomes cemented into a hardened mass that preserves a record of past life in a localized area as much as 20,000 years ago.

FUN FACT

Many cave formations are named for common household items such as draperies, needles, or soda straws. Look for them in the Museum's cave exhibit. Carlsbad Caverns formed in ancient fossil reefs similar to today's Great Barrier Reef.

Although the great continental ice sheets of North America did not extend into New Mexico, the climatic effects of the glaciations did. Isolated alpine glaciers formed in the mountains throughout the state. In the moister climate, large lakes covered the Estancia, San Augustín, Ánimas, and Tularosa valleys. The Great Salt Lake in Utah is a remnant of an even more extensive lake that covered more than 35,000 square miles during the Ice Age.

With cooler summers and more mild winters, plants and animals from the north invaded New Mexico. As the climate dried and seasons became more pronounced, the glaciers retreated along with cool, moist conditions. Life needing those conditions retreated up in elevation and to the north, while desert species moved in from the south.

FUN FACT

Bat guano deposited in caves by large colonies of free-tailed bats (*Tadarida brasiliensis*) can record changing levels of atmospheric pollution. Mercury deposits in annual layers of guano in a cave in Arizona closely followed production from a nearby copper smelter—except for a one-year delay. That was the time it took for the mercury to flow from the smokestacks to the air, to soil, to vegetation, to insects, and to the bats.

Abundant volcanoes erupted throughout New Mexico from two million years ago to 3,000 years ago. These volcanoes and volcanic flows include Mt. Taylor, the Jemez Mountains, and the Malpais near Grants and near Carrizozo, and are an important part of the landscape of modern New Mexico. By 3,000 years ago, New Mexico had taken on its modern appearance.

Fossils of the saber-toothed cat come to life in *New Mexico's Ice Age*. Photo by David Mandel, NMMNHS.

Some time before 12,000 years ago, nomadic hunters entered the New World from Siberia across the Bering Land Bridge, a wide expanse of continental shelf exposed by lowered sea levels. Direct descendants of these people, the Clovis people, soon ranged across North America. Their name comes from distinctive projectile points found near Clovis, New Mexico. The Clovis people appear to have specialized in mammoth hunting, also taking bison, horses, camels, tapirs, and other large mammals that lived in New Mexico at the time. When these large animals (*megafauna*) became extinct at the end of the last Ice Age, the Clovis culture soon disappeared.

The Clovis culture was replaced by the Folsom culture, named for distinctive fluted projectile points first found near Folsom, New Mexico. This new culture focused on hunting of the large-horned Pleistocene bison of the Great Plains.

Both the Clovis and the Folsom people had to contend with rapidly changing climates at the close of the last Ice Age. As global temperatures rose, seasonality increased, and desert vegetation expanded throughout the Southwest at the expense of woodlands and grasslands. Plant and animal communities moved north and upslope, and new plant and animal associations developed.

The large mammals on which the cultures depended

The Clovis people, named from their distinctive spear points found near Clovis, New Mexico, probably specialized in hunting mammoths (mural in *New Mexico's Ice Age* by Kent Pendleton). Photo by David Mandel, NMMNHS.

for food, particularly mammoths and bison, went extinct or were locally extirpated, perhaps in part due to hunting pressures. The cumulative effect of these environmental changes led to a shift from big game hunting to a hunter-gatherer lifestyle of people in the canyon country of the Colorado Plateau 8,000 years ago, and ushered in the Archaic period. The hunter-gatherers of the Archaic period spread throughout the Colorado Plateau, and regular trade routes connected the several hundred known archaeological sites in the Southwest.

As the drying trend gradually slowed and conditions became more stable and wetter about 4,000 years ago, cultivation of corn began in the American Southwest. First domesticated in the Tehuacán Valley of central Mexico, cultivated corn allowed people to live in one area year around. Tree ring dates from pit houses and other structures date this agricultural revolution at about 2,500 years ago. One group that took up this farming lifestyle, the Anasazi, built the great cliff dwellings of Mesa Verde, Colorado, and the impressive ancient cities of Chaco Canyon, New Mexico, and occupied an area centered on the four corners of Arizona, Utah, Colorado, and New Mexico.

By the year 1100, an estimated 5,000 people were living in and around Chaco Canyon. At least 150 satellite communities, or outliers, extended from Chaco Canyon into southern Colorado and eastern Arizona. Beginning in 1150, a series of droughts began that lasted fifty years, and efforts by the people of Chaco to bring in water via an irrigation system and to harvest timber for fuel and construction from increasing distances failed to meet the increasing demands of a growing population. The limited water and timber resources of the region could no longer support so large a population, and the people spread north to the San Juan River Valley and along the Rio Grande Valley of central New Mexico.

Living Landscapes

For many naturalists, New Mexico is truly the "Land of Enchantment" for reasons other than its cultural richness or scenic beauty. New Mexico is packed with a rich diversity of plant and animal species, due in part to its great topographic, geologic, and climatic diversity. Add to this the region's dynamic and ever-changing geologic history, and the result is a naturalist's wonderland, full of exciting discoveries.

The youthful Sandia Mountains, east of Albuquerque, are related to the formation of the Rio Grande rift. Layers of limestone deposited in a Paleozoic ocean are visible on top of Precambrian granite. The Sandias encompass five ecological life zones in about 5,000 feet of elevation. Photo by J.C. Aubele, NMMNHS.

Within the state the wide expanses of the Great Plains meet the towering Rocky Mountains, the youthful Rio Grande rift, the Basin and Range province, and the flat-lying Mesozoic rocks of the Colorado Plateau. Rivers thread their way down from alpine peaks, carving rugged canyons and mesas, wandering through desert and grassland on their journey to the sea. Isolated mountain peaks rise abruptly above broad, flat vistas. And the life is as varied as the landscape.

Drifting down the Rio Chama, time creeps by in the stillness. Creatures watch and listen from the shadows of the cottonwoods along the water's edge. Low over the water, mergansers streak by in a flash of color, winding along the river to disappear around a bend in the

The current appearance of the Rio Puerco Valley is a combination of geologic rock type and modern erosion by the Rio Puerco. Photo by L.S. Crumpler, NMMNHS.

steep canyon walls. Twittering swallows dance and dart about near the mud pockets of their cliff dwellings, feasting on a bloom of mayflies. The slap of a beaver's tail echoes off the canyon walls. Silent and patient, the colored layers of rock hold their story of millions of years' passage: of times of dinosaurs, great fields of dunes, vast inland seas, lush tropical forests, and giant volcanoes. The river slowly, relentlessly, cuts down through this history, exposing older chapters even as new layers form elsewhere.

When you began this journey through time, 4.6 billion years ago, time ticked by very quickly, in billions, then millions of years. As you approached modern-day New Mexico, the journey slowed. You have now returned to the present, but your journey through time is far from over. While the passage of time is more familiar, the processes

FUN FACT

New Mexico has both a state mineral and a state gem mineral: the state mineral is smithsonite (named after the founder of the Smithsonian); and the state gem mineral is turquoise. Turquoise has been mined in the Cerrillos area of New Mexico for more than five hundred years. The huge turquoise nugget on display in the Museum was, for many years, the centerpiece of a window in the Covered Wagon in Old Town Albuquerque. Manny Goodman, the long-time owner of the Covered Wagon, donated it to the Museum.

Wind-blown gypsum sand from a former Ice-Age lake currently is being deposited as sand dunes in the White Sands area of the Tularosa Basin near Alamogordo, New Mexico. This is the largest concentration of gypsum dunes in the world. Photo by L. S. Crumpler, NMMNHS.

that change the land and life upon it continue. Some processes appear infinitesimally slow in our familiar time frame. At times, the world appears static; in fact, it never is.

New Mexico's landscape is continually changing. Small earthquakes occur daily in New Mexico. Most of these quakes are related to the Rio Grande rift and the active volcanism in the state. Today, a constant battle exists between the forces of geology that continue to build mountains and deposit sediment and the forces of erosion that wear them away to produce mesas, valleys, and plains. Unlike many areas on Earth where the landforms only represent erosion, in modern New Mexico most landforms represent active geological processes that began in the Age of Volcanoes (Tertiary Period) and continue today.

With thousands of square miles of public lands, New Mexico is easily accessible, open to those with a yearning to explore. We hope you will explore New Mexico's natural wealth, and look upon the land from a fresh perspective: this is a dynamic, ever-changing world, shaped by on-going processes as old as Earth itself.

A brilliant white Whooping Crane feeds among Greater Sandhill Cranes in a field near Lemitar, New Mexico. Each year, hundreds of thousands of waterfowl winter in the marshes and fields along the Rio Grande in New Mexico. Photo by Dave Hafner, NMMNHS.

FUN FACT

Marshes along the Rio Grande historically were critical wintering grounds for migratory waterfowl. Agricultural development in the early 1900s drastically reduced the wetlands and bird populations: by 1939, for example, less than 1,000 Greater Sandhill Cranes survived. Thanks to a system of 400 national wildlife refuges, like Bosque del Apache FWR in New Mexico, Sandhill Cranes and other waterfowl staged a remarkable comeback: More than 25,000 cranes now winter along New Mexico's Rio Grande.

Aspen groves, like these on Mt. Taylor, occur above 9,000 feet on New Mexico mountains. Aspens provide important nesting and foraging sites for many species, and shade for slow-growing conifers. Photo by J.C. Aubele, NMMNHS.

NEW MEXICO MUSEUM OF NATURAL HISTORY & SCIENCE

The New Mexico Museum of Natural History "shall hereafter collect, preserve, study, and interpret materials representative of the natural history of the state and region and develop and maintain exhibits and programs of an educational nature for the benefit of the citizens of New Mexico and visitors to the state."
—enabling legislation, 1980

A group of aware and concerned citizens acting together for a common goal was the driving force that established the New Mexico Museum of Natural History and Science. The concern of those citizens—that New Mexico's world-renowned fossils were being taken from New Mexico to be displayed and studied at major Museums throughout the world—led to an energetic and successful effort to showcase these rare treasures in their home state. This effort began as early as ten years before the Museum opened in January 1986, and involved influential citizens, the State Legislature, residents, and New Mexico's school children.

The Museum became a reality after a public fund-raising effort to secure $2 million from private donations to match the State of New Mexico's initial investment of $8 million. That $2 million in private funds ranged from corporate sponsors to families to elementary school students who collected pennies to make the dream a reality.

The New Mexico Museum of Natural History—the "and Science" part of the name was added in 1990—opened on January 12, 1986 after a two-year construction project to build the Museum in Old Town Albuquerque on land donated by the Public Service Company of New Mexico.

It is located at 1801 Mountain Road NW.

The Museum that opened in 1986 with four permanent exhibit halls is much different than the Museum that exists today. The Museum's most comprehensive permanent exhibit hall—*New Mexico's Seacoast*—opened in July of 1988, more than two years after the Museum first opened to the public. With its living forest, one-of-a-kind Evolator time machine, and two floors of exhibits, the *Seacoast* Hall is one of the Museum's most dramatic. Other permanent exhibit halls are *Origins*, (the Precambrian through the Paleozoic Era), *Age of Giants* (Jurassic Period), *Age of Volcanoes, and Evolving Grasslands* (Tertiary Period), and *New Mexico's Ice Age* (Pleistocene Epoch). In the near future, the Museum plans to add two new permanent exhibit halls and upgrade the *Seacoast* Hall. The new halls will feature New Mexico's Triassic Period (*Dawn of the Dinosaurs*), focusing on the processes of evolution, and current day New Mexico (*Living Landscapes*), focusing on the interplay of climate, topography, geological processes, and biological evolution that continue to shape our world.

(Above) The Museum atrium in the middle stage of construction. NMMNHS photo.

(Below) The Museum Atrium as it looks today. Photo by David Mandel, NMMNHS.

(Opposite) Two life-sized bronze reproductions of Cretaceous-age dinosaurs by artist Dave Thomas greet visitors at the entrance to the Museum. The horned, herbivorous *Pentaceratops* is nicknamed "Spike" and the carnivorous *Albertosaurus* is nicknamed "Alberta." Photo by David Mandel, NMMNHS.

Since 1983, Museum scientists, educators, and exhibit designers and fabricators cooperatively have produced the permanent exhibits—*TimeTracks: A Journey Through the Natural World*—as well as smaller exhibits that travel throughout New Mexico and surrounding states. In addition, the Exhibits

Department periodically mounts major traveling exhibits from other institutions in the Museum's large changing exhibit hall.

The original large-format theater opened in August 1989, and it was the first flat-screen Iwerks Theater in the country. Ten years later, in October 1999, a much larger version—the Extreme Screen DynaTheater—opened, one part of an ambitious construction project that also included expanded store space, new education classrooms, a new entrance, and the LodeStar Astronomy Center.

Since its opening in 1986, the Museum has welcomed more than 4.5 million visitors from around the world. It is the most-visited Museum in the state and a well-respected institution with a collection of more than 150,000 natural objects ranging from the world's oldest mammal fossil to the world's longest dinosaur. It has received accreditation and subsequent accreditation from the American Association of Museums.

The Museum has presented more than one hundred temporary science and art exhibits, ranging in subject matter from radar images of Venus' surface to Chinese dinosaurs, Fabergé eggs, and a look at sports science.

This variety of temporary exhibits allows the Museum to explore subjects not included in its permanent exhibits and also to reach out to new audiences. Since the opening of its 6,000-square-foot Traveling Exhibits Hall in 1997, the Museum has been able to bring in larger, more interactive exhibits from national and international venues such as *Spiders* from the Smithsonian Institution and *Sport* from the Ontario Science Center.

The Museum also develops its own exhibits and has toured them throughout New Mexico and surrounding states. *The Night Shift: A Closer Look at Bats*, *Fire: A Natural Process*, and *Wild Plants of the Pueblo Province* are three of the exhibits that have been produced in cooperation with New Mexico scientists and naturalists.

A block of rock from a *Coelophysis* quarry near Ghost Ranch, New Mexico, encased in protective plaster, was moved into the Museum in 1985. Today, it is being slowly and painstakingly prepared in the Museum's *FossilWorks* exhibit. NMMNHS photo.

The Museum's Science Division conducts collections-based research in paleontology, geology, and biology. Supported by outside grant support, cooperative relationships with federal agencies, and an enthusiastic corps of volunteers, the Museum's curators have developed important collections in the natural history sciences that serve as the foundation for the public exhibits as well as research.

Largely due to its growing collections, in 1993 the Museum purchased a neighboring 18,500-square-foot building on 3.3 acres of land. Renovation of the building as the Research and Education Complex was completed in 1996.

The natural treasures preserved in the Museum's collections define New Mexico's natural history and heritage, and provide the tangible evidence of a wondrous past. Preservation of our natural heritage is vitally important, yet the Museum is more than a showcase for New Mexico's natural treasures. The Museum uses these objects to convey a dynamic and exciting view of our past, a history where all things are always changing.

The Museum's Education Department administers educational programs within and outside of the Museum that actively involve teachers and school children in the natural history sciences. These programs range from summer programs that explore the natural world and training of docents to interpret the Museum's exhibits to teacher workshops, an educational television series, and a bilingual science program that seeks to enhance natural history education among Hispanic families and school children. The Department also manages an off-site, outdoor educational facility, the *Sandia Mountain Natural History Center*, in the Sandia Mountains east of Albuquerque.

The Lodestar Astronomy Center is designed to engage, educate, and entertain visitors of all ages with the excitement of space science. The Center features a 55-foot diameter planetarium, an observatory with a 16-inch telescope, exhibits on the scale and structure of the universe, and the StarWorks Astronomy Store.

The New Mexico Museum of Natural History Foundation supports the Museum through various fund raising efforts, entrepreneurial activities, marketing, management of grants to the Museum, and development of general Museum support.

A Division of the State of New Mexico's Office of Cultural Affairs, the Museum—which is accessible to persons with disabilities—continues to nurture an individual's sense of wonder of the natural world, particularly that of New Mexico and the Southwest.

The rock wall in the Museum's Atrium represents rock layers from central New Mexico. As visitors follow *TimeTracks* through the Museum, each of the time periods and environments represented by the rock wall are displayed in a separate hall. Photo by David Mandel, NMMNHS.

Lockheed Martin Extreme Screen DynaTheater

In August 1989, the Museum opened the first Iwerks large format theater in the country. In October 1999, the Museum opened the new Lockheed Martin Extreme Screen DynaTheater, which brings the latest in technology to the giant screen in Albuquerque.

The new Extreme Screen DynaTheater features a giant screen measuring 65 feet wide by 49 feet high, which is three times the area of the original screen, and stadium-style seating for 284 people compared to the original theater's 96 seats.

The theater has several significant differences in the systems that make up a large format theater as compared to traditional commercial cinemas, which show 35-mm film. The new Extreme Screen DynaTheater utilizes 8/70 film, twice the size of 35-mm film.

The larger Iwerks' projection system utilizes "Iwerks Linear Loop" technology, one of the first major advancements in projection in three decades, and delivers the brightest, steadiest images for 8/70 3D and 2D formats. This

Sea kayaks await *Explorations!* participants on a whale-watching trip to Bahia Magdalena, Baja California Sur, Mexico. *Explorations!* trips often are related to films presented at the Museum's Extreme Screen DynaTheater. Photo by Dave Hafner, NMMNHS.

projection technology, combined with a six-story screen that is three times the size of that located at the original Iwerks theater at the Museum, provides viewers with an opportunity to experience a truly immersive entertainment adventure.

The advancement in projection technology pioneered by Iwerks, coupled with increased interest and success in the development of commercial and educational large format film, enhances the Museum's ability to provide quality content in a unique viewing environment.

Visitors to the Museum's Extreme Screen DynaTheater see breathtaking views of the wonders of nature such as this eruption of Volcán Arenal in northern Costa Rica. The Museum's travel program, *Explorations!*, has taken participants to Costa Rica twice. Photo by Dave Hafner, NMMNHS.

The new Extreme Screen DynaTheater adds to the Museum's position as a key member of the Iwerks family, which includes one-third of all the giant screen theaters in the world and two-thirds of all the 8/70 giant screen theaters worldwide. Key features of the new system include system-ready expansion to 3D and multi-format projection, which will allow for the exhibition of 35-mm as well as 8/70 films.

The Extreme Screen DynaTheater provides optimal seating for visitors with disabilities. An LED-projected captioning system for the hearing impaired and descriptive services for the visually impaired is in place. Wheelchair access is available in the front and rear of the theater in optimal viewing areas.

These theater improvements were made possible in large part by a $1 million gift from Lockheed Martin/Sandia National Laboratories and other Capital Campaign donors as well as the State of New Mexico. The New Mexico Museum of Natural History Foundation manages and runs the theater.

LodeStar Astronomy Center

The Museum is home to the 18,000-square-foot LodeStar Astronomy Center. Operated by the University of New Mexico, the astronomy center features a multimedia planetarium, spectacular exhibits, a motion-simulation theater, an astronomy-themed retail store, and an observatory.

The astronomy center is designed to engage, educate, and entertain visitors of all ages with the excitement of astronomy. Visitors will learn basic scientific principles as they journey from Earth through our solar system, the Milky Way galaxy, our local group of galaxies, huge clusters of galaxies, even larger scaled structures, and ultimately, the entire Universe.

Exhibits on the first floor describe the scale and structure of the Universe. These provide opportunities to learn more about the objects and patterns of the Universe through hands-on exhibits, interactive multimedia devices, models, displays, projects, and programs. Besides offering wonderful learning opportunities, these exhibits also provide an introduction to topics explored more deeply in the planetarium and motion-simulator experiences.

Planetarium

Inside the 55-foot planetarium, planetarium shows incorporate the latest technology to reveal the beauty and mystery of the Universe. The raked seats and domed screen convey a sense of total immersion and movement during programs while also creating a unique space for lectures and special events.

Construction workers settle the LodeStar Planetarium dome onto the new building in September 1999. The new LodeStar Astronomy Center opened at the Museum in January 2000. Photo by John Arnold, NMMNHS.

What can you expect when you recline in one of the theater's 150 seats? You will find journeys through time, space, and the natural world. With a spectacular full-dome video system and a digital star projector, the theater far exceeds the capabilities of a classic planetarium—it is more of a time-space machine. The Planetarium's cutting-edge tools project full-color, high-resolution images that move across the entire 4,750-square-foot screen. Visitors can see the world through an insect's eye, examine a living cell, explore the double helix of DNA, or visualize mathematical functions and constellations.

Observatory

The observatory houses a 16-inch Schmidt-Cassegrain telescope. The sun can be viewed through special filters during daytime visits. The stars, planets, and other celestial objects are visible during nighttime programs and special events. The telescope also augments the planetarium by providing real-time astronomical images for projection on the dome. Light pollution in

The LodeStar Astronomy Center includes an observatory, a Planetarium, and exhibits, all of which help visitors explore the cosmos. Photo by David Mandel, NMMNHS.

Albuquerque's night skies limits the capabilities of the observatory, but provides an opportunity to help visitors understand good lighting design and its benefits.

StarWorks Astronomy Store

The LodeStar Astronomy Center features StarWorks, an astronomy-themed retail store that extends LodeStar's theater shows, exhibits, and science education programs. The store features educational books, posters, puzzles, toys, games, charts, minerals, science kits, globes, computer software, and telescopes. The store allows visitors of all ages to take home some of the fun and learning of the astronomy center and provides a place to shop for educational gifts.

Funding

A $15.8 million federal grant to the University of New Mexico's Institute for Astrophysics and $12 million in matching funds from the State of New Mexico funded the LodeStar Project. LodeStar comprises numerous educational programs and sites that work together to improve the public's knowledge of and interest in science, while providing school children with educational opportunities that are otherwise unavailable. The LodeStar Astronomy Center represents a $10 million investment by the project, and Enchanted Skies Park, a future phase of the project located west of Albuquerque, will extend the project's goals to dark-sky observing for researchers, amateurs, and the general public.

Research and Collections

Volunteers trained by the Museum's preparators prepare fossil specimens and answer visitors' questions in the *FossilWorks* exhibit. Photo by David Mandel, NMMNHS.

Natural science collections are documents of change, the fragmentary archive on which we base our knowledge of the natural world. Encoded within these collections are the past of our planet, the ledger of our interaction with nature, and the record we decipher to ensure a future for our own and other species.
—Executive Summary, *Preserving Natural Science Collections: Chronicle of Our Environmental Heritage*

Natural history collections are defined as organized collections founded on biological and geological specimens, together with associated information and

Museum staff and volunteers collecting skin impression of a 70-million-year-old fossil hadrosaur near Deming, southern New Mexico. Detail: close-up of skin impression. Photos by Spencer Lucas, NMMNHS.

expertise. The permanent collections of the New Mexico Museum of Natural History and Science—indeed, of any natural history museum—are the tangible foundation for all of its research, education, and exhibit programs. They are part of larger, global collections built over the past three hundred years and held in museums and other institutions that are responsible for facilitating access to and use of such resources, and for their care and development for the benefit of society. Importantly, these collections are nonrenewable riches that document the record of our environmental past and our level of understanding of our current world. They are the ultimate resources upon which science must depend as increasingly refined analytical tools are developed, even as species, habitats, and fossil sites disappear.

The immediate scientific uses of natural history collections—scholarly publications—initially may appear to be of little consequence to the average museum visitor. Yet these studies, usually reported in the abstruse and esoteric lexicon of formal science, have yielded notable benefits to society at large. For example, collections of living microorganisms and plants have been used to derive medicinal drugs that have saved countless lives and relieved suffering worldwide. By documenting the occurrence of plants and animals, biological collections provide the basis for environmental managers to make informed choices as to preferred sites of development or preservation. As we learn more about past environmental and geological changes and their impacts on past life forms recorded in the fossil record, we may more accurately predict the consequences of such future change. Moreover, the display of natural history specimens, and the interpretation of their temporal and environmental setting, promotes greater public understanding of environmental issues and of the process of science itself. In this way, these collections are an expression of our need to understand the world, an enriching human pursuit in its own right.

Emphasis on Paleontology

New Mexico has long been famous for its fossil riches. For example, paleontologists from the American Museum of Natural History in New York conducted numerous fossil-collecting expeditions to New Mexico from 1890 to 1960, resulting in large collections and impressive public exhibitions. The initial impetus for founding the Museum was to preserve for the state's citizens examples of these fossil treasures. This original focus on paleontology was expanded by the founding legislation to embrace the

Virtually complete fossil skeleton of *Typothorax*, an aetosaur of New Mexico's Triassic. These vegetarians were heavily encased in armor, likely as protection against the active predators, the phytosaurs. Photo by Spencer Lucas, NMMNHS.

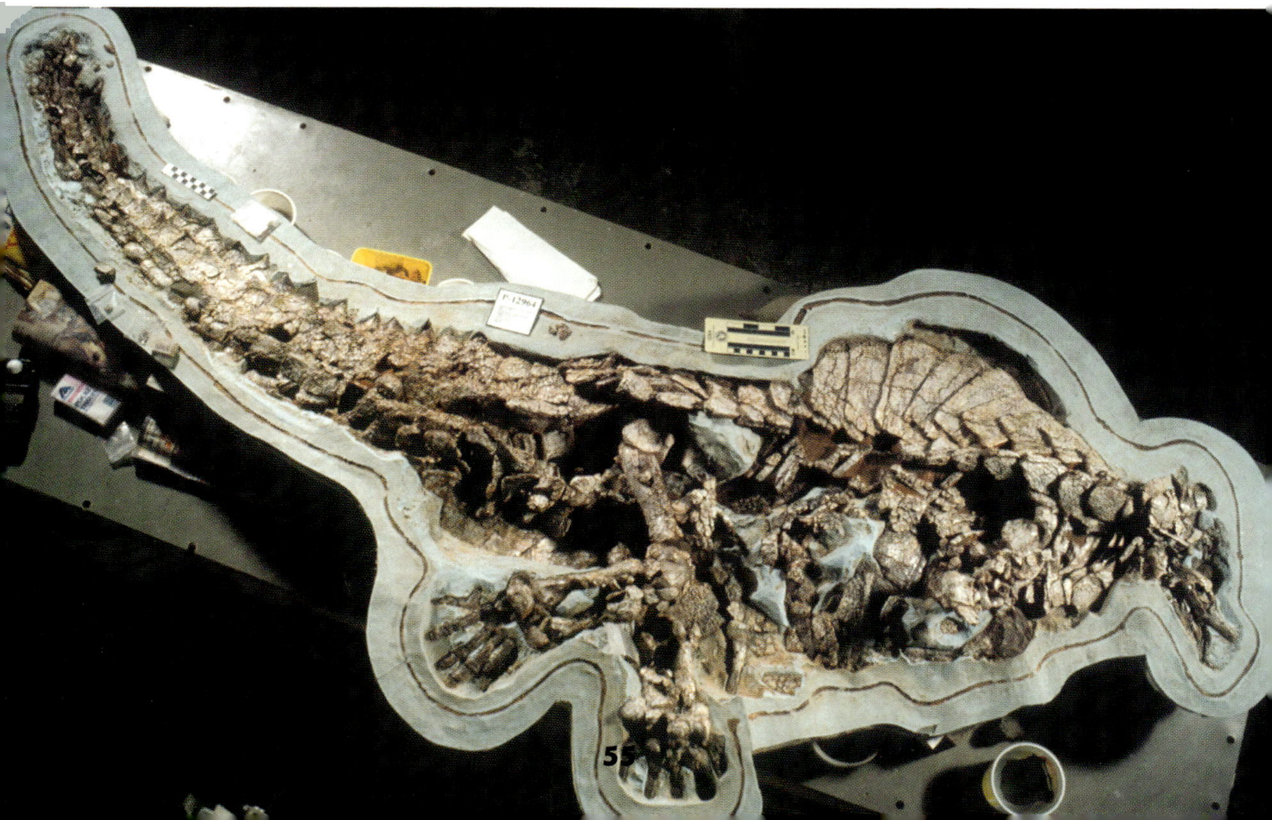

biological as well the geological sciences, and to "collect, preserve, and interpret" the State's and region's natural history riches for the Museum's public.

The Museum retains its geographic focus on the Southwest (southwestern United States and northern Mexico), with a particular emphasis on New Mexico. In acknowledgement of the existing natural history collections in the state, particularly at the major universities, the Museum chose to emphasize disciplines that would be complementary to, rather than competing with, those collections. The need for development of a regional paleontological collection was judged to be paramount, and has resulted in the development of a collection of some 125,000 specimens. Built by staff fieldwork and the inclusion of two collections from other state institutions, this collection satisfies the criteria of a Regional Systematic Research Resource Center proposed by the Association of Systematic Collections in 1991. Science staff share responsibility for the conservation and preservation of collections and associated information with the Registration Department; the Museum's Board of Trustees maintains fiduciary control of the collections.

The paleontological collections of the Museum have benefited from annual support through cooperative agreements with federal agencies, particularly the U.S. Bureau of Land Management. Agreements with these agencies and the New Mexico State Land Office also designate the Museum as the repository for fossils not collected under permits by other research institutions.

Reconstruction of the internal chambers of the crest of a Cretaceous hadrosaur, *Parasaurolophus tubicen*, based on computer tomographic and radiograph data. NMMNHS image.

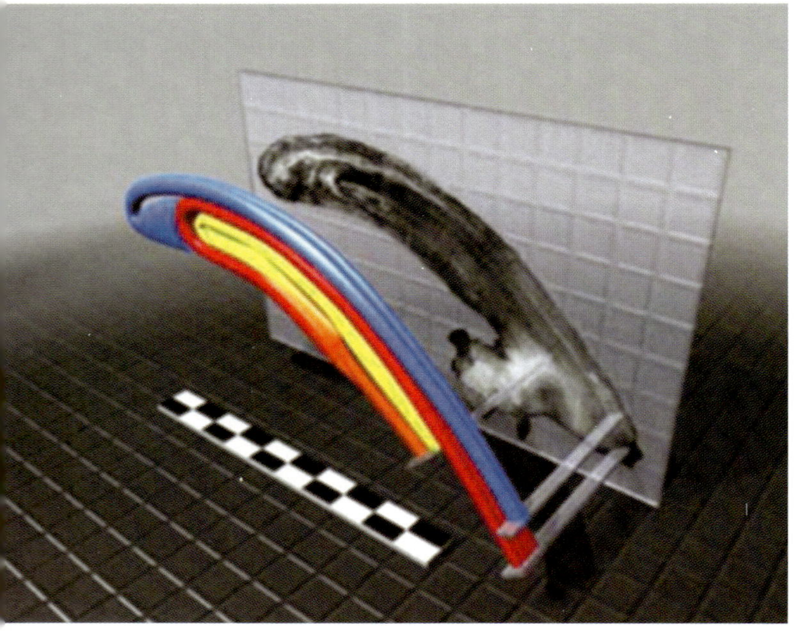

Research

Since its public opening in 1986, the Museum's Science Division has built a collections-based research program with a focus that is best defined as historical biogeography: the study of the changing geographic distribution of life on land. Supported by state funding augmented by research grants from a variety of foundations, a cadre of staff curators in association with a larger group of research curators, adjunct curators, and colleagues at other institutions studies the evolution of a constantly changing world. Whether it involves the description of a new form of dinosaur from the San Juan Basin, the dynamics of a volcanic eruption, the effects of past environmental oscillations on the local scene, or past distributional changes preserved in the genetic record of living populations, the research staff of the Museum works to further our understanding of past, current, and future change, while continuing to build the tangible record upon which this understanding is based.

Genetic studies of the antelope ground squirrel conducted by Museum scientists provide an insight into the evolution of other desert life in southwestern North America, including that of the Baja California peninsula and islands in the Sea of Cortéz. Photo by Dave Hafner, NMMNHS.

Major research efforts of the curatorial staff have been supported by grants from the National Science Foundation, the National Geographic Society, the National Aeronautics and Space Administration, and regional institutions. Collaborative grants conducted with colleagues at other institutions have forged lasting relationships between the Museum and similar research programs at the regional, national, and international levels.

Active participation of the curatorial staff in regional, national, and international scientific organizations and symposia, as well as hosting of scientific symposia and conferences at the Museum, has developed a strong reputation for the Museum. This scientific reputation was further enhanced by initiation of the *Bulletin of the New Mexico Museum of Natural History and Science* in 1992. This scientific bulletin is issued at least once annually, focusing on topics in natural history ranging from *The Nonmarine Triassic* (no. 3; 1993) and *Early Permian Footprints and Facies* (no. 6; 1995) to *Land Snails of New Mexico* (no. 10; 1997) and *Dinosaurs of New Mexico* (no. 17; 2000).

Education

Besides the immediate tasks of collecting specimens from the field, maintaining collections, and conducting ensuing scientific research, staff of the Science Division has major responsibilities in development of exhibitions for public education (shared with the Exhibits and Education departments) and participation in the public educational programs of the Museum (developed by the Education Department). In contrast to many natural history museums, the scientific staff enthusiastically participate in the Museum's public programs, considering this role to be of vital importance in bridging the gap between academic research and the Museum's public. The Museum shares the objective of enhancing communication between scientists and the general public with the New Mexico Academy of Science, which was founded in 1902 and has been formally housed at the Museum since 1995.

The Science Division provides a diverse array of opportunities for involvement by students and volunteers. Through their adjunct appointments at universities, curators direct graduate students who actively participate in collections and research activities at the Museum while taking coursework at the universities. Grant-supported research programs have included graduate and undergraduate students from universities in the United States and Mexico, as well as high-school students in gifted programs at local high schools. These students obtain first-hand experience in field and laboratory research as well as collections management. Volunteers provide much appreciated assistance in fieldwork and collections, and in return learn about the research in which they are participating. Volunteers often provide much of the labor involved in major fossil quarries, conducted under the supervision of curators, their students, and other more-experienced volunteers. One program, supervised by the chief preparator and conducted with the assistance of the Friends of Paleontology (affiliated with the NMMNH Foundation), trains volunteers in fossil preparation and interpretation. These trained volunteers then work on fossils either in the preparation laboratory or in FossilWorks, a public exhibition that showcases the processes of fossil preparation. Curators also teach regular classes in the docent-training program, which provides a scientific background for volunteer exhibit interpreters, and occasionally offer public lectures and university classes.

Educational Programs

The Museum offers educational activities and resources throughout the state for teachers, students of all grade levels, parents, families, and the general public. Natural History and Science museums are highly effective sites for science education. In fact, education is central to the mission of the Museum. Visitors of all backgrounds and all ages can learn something new in an informal (non-classroom) setting. Museum specimens provide a physical link, for many visitors, to otherwise abstract science topics; the objects in this Museum bring the natural history of New Mexico to Museum visitors in a fun and informative way. New scientific research findings by the curators of the Science Division are used to keep exhibits and educational programming current. Learning in a museum setting includes hands-on participation and exploration. The Museum teaches the public through its permanent and short-term exhibits and through educational programming targeted to specific age groups. The Museum also acts as a science content provider by presenting science topics, creating curricula, and loaning teaching specimens to classroom teachers, who would be unable to easily acquire such items.

Students explore the youthful Albuquerque volcanoes as part of the "Very Cool Volcanoes" educational program. Photo by Kate Dowe, NMMNHS.

Museum volunteers experience the natural history of New Mexico and the southwest during monthly day trips and annual week-long field trips led by the Museum's education and science divisions. Photo by Marilyn Ortman, NMMNHS.

Yearlong and Life-long Learning

The Education Division of the Museum makes supplemental learning and teaching materials, including kits and curriculum, available to teachers and parents. Our hands-on educational room in the Museum, the *Naturalist Center*, is for "children" of all ages. The *Museum Lecture Series* is designed to bring scientists and their work to interested non-scientist adults and older children. Teenagers learn public presentation skills and science while they volunteer during the summer as Junior Docents. Also during the summer, the *Young Explorer's Program* offers weeklong science day camps for children from pre-school through elementary grades. Middle school and high school students statewide can participate in ecological research projects as part of the *Summer Ecology Research Program*.

A large and active group of volunteers of all ages and backgrounds who work in all phases of the Museum operation, from behind the scenes to visitor education, serve the Museum. A full-time Volunteer Coordinator who is a member of the Museum Education Division manages the *Volunteer Program*. The Museum staff provides extensive training and education programs for volunteers, including specialized training, monthly field trips and continuing education classes. Each year, volunteers contribute more than 33,000 hours.

Special Interest Programs

The Museum offers several extensive programs that include curricula for students, professional development workshops for teachers, and parent components. These programs include: *Proyecto Futuro*, a bilingual science program for teachers and parents of elementary to middle school students; *Topo New Mexico*, a project for 8th–12th grade students involving the production of a scale map of the state by students in schools throughout the state; and *Way Out West: The Electronic Bridge*, our nationally distributed distance learning television program about ecosystems and ecological research for middle and high school students.

By agreement with Albuquerque Public Schools, the Museum operates the *Sandia Mountain Natural History Center* in the Sandia Mountains. The Center is an outdoor environmental education facility visited by about 8,000 5th grade students and teachers each year. During the winter, instructors from the Center travel to communities throughout New Mexico with an outreach program.

Both the *Sandia Mountain Natural History Center Program* and *Way Out West* have won awards from the EPA for Excellence in Environmental Education.

Special Events

A variety of special events, science presentations, and hands-on family learning activities happen throughout the year at the Museum. These include teacher workshops, our annual *Teacher Open House*, science-related performances, and presentations by our Magician-in-Residence, a trained educator who uses magic to teach about science. Special events are frequently conducted in conjunction with other institutions.

The Museum Education Division and Museum Volunteers run two special annual events. *Spike's Ride*, a family fun bike ride and chile festival in September, is organized jointly by the Museum and REI. *Dino Dash*, organized by the Museum volunteers, is one of the largest sanctioned races for runners in New Mexico. Proceeds raised from both events are used to fund a variety of educational programs and other needs at the Museum.

Students in the Museum's "Time Trekkers" program climb down from the summit of Mt. Taylor, a two-million-year-old volcano in western New Mexico. Photo by Helen Haskell, NMMNHS.

New Mexico Museum of Natural History Foundation

The New Mexico Museum of Natural History Foundation, Inc. is a private, non-profit corporation whose sole purpose is to further the cultural and scientific programs and purposes of the New Mexico Museum of Natural History and Science. The Foundation's objectives are to receive, raise and administer funds for the benefit of the Museum and to provide the Museum with a widespread base of support.

A blue-footed booby shows off his striking feet to *Explorations!* guests on an expedition to the Galápagos Islands of Ecuador. Photo by Dave Hafner, NMMNHS.

The Foundation provides many services that support the Museum's research and outreach mission. The Foundation manages all grants received by Museum scientists and educators; develops valuable radio, outdoor, print and television promotion for the Museum; and administers the Museum's travel programs.

Explorations! is the Museum's principal travel program, emphasizing education, discovery and adventure for adults. Museum curators lead public participants on natural history excursions throughout New Mexico, the Southwest, and the world. *Explorations!* has offered trips to the Galápagos Islands, China, Kenya, Egypt, Costa Rica, Belize, and Mexico's Yucatan and Baja California peninsulas, as well as more local trips that include rafting along the Colorado River, Chama River, and Rio Grande; caving in the Guadalupe cave system; bat-netting; and fossil hunting. Another regular program, the *Southwest Institute*, provides an outstanding experience for those who want to travel throughout the greater Southwest and learn of its exciting cultural, historical, and natural diversity.

Foundation Enterprises

Through a series of formal agreements with the Museum, the Foundation operates several major revenue-generating enterprises, which include the Camp-In program, the NatureWorks Discovery Store, and the Extreme Screen DynaTheater. The *Sleep with the Dinosaurs* Camp-In program is an exciting educational activity for children in grades 1–6 that takes campers on a 4.6-billion-year journey in an overnight Museum visit. Within the NatureWorks Discovery Store visitors will find a wide selection of merchandise pertaining to the Museum's permanent and special exhibits, such as educational games, toys, and books, including the Museum's *Bulletin*.

Development Activities

Corporations, foundations, and individuals throughout New Mexico continue to demonstrate financial commitment to the Museum, and the Foundation serves as the liaison between the Museum and donors. Gifts and benefits can be individually tailored to a donor's financial situation while providing a significant legacy to New Mexico's future. Corporate sponsorships of various events and exhibits prove to be beneficial to both businesses and the Museum. An Endowment Program and Planned Giving opportunities are also available for donors who seek to offer long-term support.

Funds raised by special events benefit the Foundation and its efforts to support programs at the Museum. The annual gala black-tie fund-raiser "Chocolate Fantasy" has become one of the premier parties in New Mexico. Decadent chocolate creations by some of New Mexico's finest chefs highlight the evening. A silent auction helps to raise funds that ultimately go to further the cultural and scientific programs of the Museum.

The Museum Foundation manages NatureWorks, the fun and discovery store. Photo by David Mandel, NMMNHS.

Membership

The Museum offers a variety of membership options to its patrons; the Foundation manages the membership program. Basic membership levels are available for families, grandparents, seniors, individuals, and students.

Benefits of membership include: free Museum admission throughout the membership year; free admission to more than two hundred other Museums around the country; free passes to the Extreme Screen DynaTheater; a subscription to the *TimeTracks* newsletter that highlights information about exhibits, events, and programs happening at the Museum; a discount at NatureWorks Discovery Store; special discounts on programs at the Museum, such as lectures, classes, field trips, and travel programs; and invitations to special events, such as a Museum Open House and film premieres.

The Foundation also encourages contributing levels of membership that include a donation that supports the exhibits, scientific, and educational programs at the Museum.

The purpose of the Coelophysis Society, the most prestigious level of membership, is to promote a lasting relationship between donors and the Museum. In return for their support and contributions to the Museum, Society members may attend private lectures with the curators of the Museum, exclusive field trips, and previews of films and exhibits.

The Foundation maintains a Membership Desk at the entrance to the Museum, which provides information to both active and potential members. Museum guests are invited to inquire about Museum membership programs, enroll in activities sponsored by the Foundation, and purchase memberships for their families and friends.